Investigating Rocks

Natalie Lunis and Nancy White

Newbridge Educational Publishing, LLC
New York

Table of Contents

Investigating Rocks
ISBN: 1-58273-082-2

Written by Natalie Lunis and Nancy White
Edited by Jennifer Prescott and Joanne Allgor
Designed by Frances Schlesinger
Production Manager: Michael Miller

Newbridge Educational Publishing
333 East 38th Street, New York, NY 10016
www.newbridgeonline.com

Photo Credits:
Cover: Chromosohm/Sohm/The Stock Market; Page 1: David D. Keaton/The Stock Market; Page 3: Chip Henderson/Index
Stock Photography, Inc.; Page 4: Jeff Gnass/The Stock Market; Page 5: Eric Meola/The Image Bank; Page 6: (top) Richard
Longseth/DRK Photo, (inset, bottom) Alan Kearney/FPG International; Page 7: Ed Lallo/The Picture Cube, Inc.; Page 8:
(from top to bottom) Andrew J. Martinez/Photo Researchers, Inc., E.R. Degginger/Color-Pic, Inc., E.R. Degginger/Color-Pic,
Inc., E.R. Degginger/Bruce Coleman, Inc.; Page 9: (clockwise from top left) E.R. Degginger/Color-Pic, Inc., E.R. Degginger/
Color-Pic, Inc., Biophoto Associates/Photo Researchers, Inc., Joyce Photographics/Photo Researchers, Inc.; Page 10: (top)
Mark E. Gibson/Mark Gibson, (bottom, inset) Art Wolfe/Tony Stone Images; Page 11: Bruce Clendenning/Bruce Coleman,
Inc.; Page 12: Gene Peach/Tony Stone Images; Page 13: (clockwise from top) Tom Bean/DRK Photo, Omniphoto/Index
Stock Photography, Inc., Lee Foster/FPG International; Page 14: Richard Elliott/Tony Stone Images, (inset) E.R. Degginger/
Color-Pic, Inc.; Page 15: (clockwise from top) Buddy Mays/FPG International, (inset) Stephen Trimble/DRK Photo, Alain
Morvan/Liaison Agency; Page 15: E.R. Degginger/Bruce Coleman, Inc., Lester Lefkowitz/The Stock Market, (inset) E.R.
Degginger/Color-Pic, Inc.; Page 16: (top) Pat O'Hara/DRK Photo, (bottom) Richard Price/FPG International; Page 17:
(clockwise from top) Tom Trill/Tony Stone Images, Brett Palmer/The Stock Market, David D. Keaton/The Stock Market;
Page 18: Mark E. Gibson/Mark Gibson

Illustration credit: Page 5: (inset) Jared Schneiderman Design

10 9 8 7 6 5 4

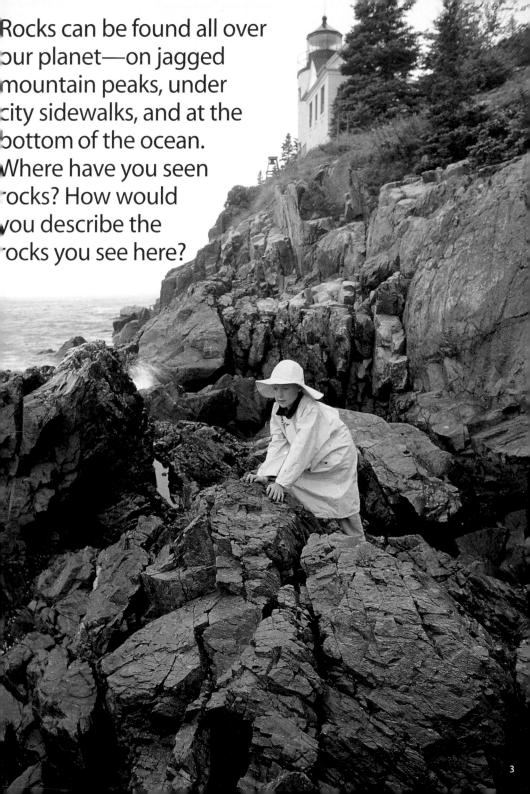

Rocks can be found all over our planet—on jagged mountain peaks, under city sidewalks, and at the bottom of the ocean. Where have you seen rocks? How would you describe the rocks you see here?

3

The Earth's Crust

Most rocks are pieces of the Earth's **crust**, its hard and rocky outer layer. Much of the crust is covered by water, ice, or soil. But sometimes parts are uncovered, or *exposed*, like this giant patch of rock.

Exposed patches of the Earth's crust are called **outcroppings**.

There is also rock beneath the Earth's crust. Sometimes cracks form in the crust, allowing melted rock from deeper layers to reach the surface.

Study this diagram. Does the Earth get hotter or colder toward its center?

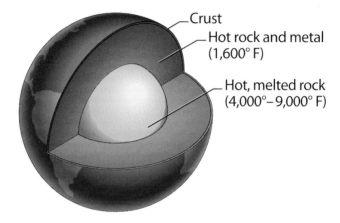

Crust

Hot rock and metal (1,600° F)

Hot, melted rock (4,000°–9,000° F)

Hot, melted rock called **lava** is erupting from this volcano. Where do you think lava comes from?

Rocks come in many shapes and sizes. They can be huge, like these boulders, or small, like the colorful pebbles below.

A scientist who studies rocks is called a **geologist**. Geologists use many tools to observe and *classify* rocks. Name the tools these young geologists are using to study rocks they have found.

One of the tools a geologist uses is called a **field guide**. In a field guide on rocks, you'll find pictures and descriptions like the ones below.

Sandstone (SAND-stone) is a grainy rock that is often reddish in color.

Basalt (buh-SALT) is a hard, smooth rock that often has holes in it. It may be gray, green, or black.

Granite (GRAN-it) is a rough, grayish rock that is usually speckled with white and pink.

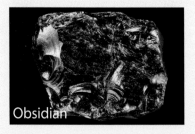

Obsidian (ub-SIH-dee-un) is a glasslike rock that is black and shiny. It usually has sharp edges.

Imagine that you've collected these four rocks. List all the colors that you see. Are the rocks shiny or dull? Do they seem smooth or rough?

Record their features. Now use the pictures and descriptions on page 8 to identify each one.

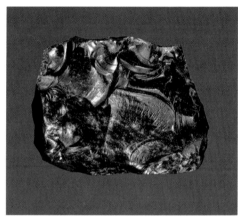

Erosion

Why are rocks different shapes and sizes? They are broken down and shaped by the forces of water and wind. This process is called **erosion**. The beach here has been eroded by the constant waves.

These pebbles have been worn smooth by huge sheets of ice grinding over them. How long do you think this took?

These rocks were also shaped by wind and weather over time. The wind sweeps up grains of sand and whips them against the rocks. The sand, carried by the wind, has carved the rocks into strange and beautiful shapes.

Predict what might happen to this sandstone shape after many years.

Using Rocks

With the help of tools, people use rock for many useful purposes. The labeled material in this photo comes from rock.

Iron

Bricks

Cobblestones

Which things in these three photos are made from rock?

Can you think of other places you might see rocks shaped by people?

People can also create art and beautiful objects out of rock. This Egyptian mask is made with **gold**. The pottery is made from **clay**. The statue is made of **marble**. The necklace is made with **diamonds**.

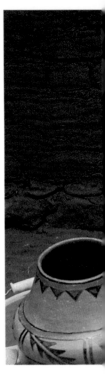

Compare each work of art with the material from which it was made.

How are they alike? How are they different?

Classifying Shapes

Classify these photos of rocks. Which show rock in its natural state? Describe the forces—such as wind, water, or ice—that might have caused the shape and size of each.

Which photos show objects that people have made from rocks? Describe how you think each object was made.

Our planet is covered with rocks. You can start investigating right outside your door. What kinds of rocks do you think you'll find?

Glossary

basalt (buh-SALT): A smooth, gray, green, or black rock that often has holes in it.

crust (KRUST): The hard, rocky outer layer of the Earth.

erosion (ee-RO-zhun): The wearing down of rocks by natural forces, such as wind, water, weather, or glacial ice.

field guide (FEELD GIDE): A book that provides pictures and descriptions of different types of things, such as rocks, which can be helpful in identifying and classifying them.

geologist (jee-OL-uh-jist): A scientist who studies rocks and the Earth.

granite (GRAN-it): A grayish rock, often speckled with white and pink.

lava (LAH-vuh): Hot, melted rock that usually emerges from a volcano or a crack in the Earth.

obsidian (ub-SIH-dee-un): A usually sharp-edged, glasslike rock that is black and shiny.

outcropping (OUT-krop-ing): An exposed area of the Earth's crust.

sandstone (SAND-stone): A grainy rock that is often reddish in color.

Index